1 MONTH OF
FREE
READING

at
www.ForgottenBooks.com

By purchasing this book you are eligible for one month membership to ForgottenBooks.com, giving you unlimited access to our entire collection of over 1,000,000 titles via our web site and mobile apps.

To claim your free month visit:

www.forgottenbooks.com/free1300467

ISBN 978-0-428-69652-8
PIBN 11300467

**CIHM
Microfiche
Series
(Monographs)**

**ICMH
Collection de
microfiches
(monographies)**

Canadian Institute for Historical Microreproductions / Institut canadien de microreproductions historiques

Technical and Bibliographic Notes / Notes techniques et bibliographiques

The Institute has attempted to obtain the best original copy available for filming. Features of this copy which may be bibliographically unique, which may alter any of the images in the reproduction, or which may significantly change the usual method of filming are checked below.

L'Institut a microfilmé le meilleur exemplaire qu'il lui a été possible de se procurer. Les détails de cet exemplaire qui sont peut-être uniques du point de vue bibliographique, qui peuvent modifier une image reproduite, ou qui peuvent exiger une modification dans la méthode normale de filmage sont indiqués ci-dessous.

- [✓] Coloured covers /
  Couverture de couleur

- [ ] Covers damaged /
  Couverture endommagée

- [ ] Covers restored and/or laminated /
  Couverture restaurée et/ou pelliculée

- [ ] Cover title missing / Le titre de couverture manque

- [ ] Coloured maps / Cartes géographiques en couleur

- [ ] Coloured ink (i.e. other than blue or black) /
  Encre de couleur (i.e. autre que bleue ou noire)

- [ ] Coloured plates and/or illustrations /
  Planches et/ou illustrations en couleur

- [ ] Bound with other material /
  Relié avec d'autres documents

- [ ] Only edition available /
  Seule édition disponible

- [ ] Tight binding may cause shadows or distortion along interior margin / La reliure serrée peut causer de l'ombre ou de la distorsion le long de la marge intérieure.

- [ ] Blank leaves added during restorations may appear within the text. Whenever possible, these have been omitted from filming / Il se peut que certaines pages blanches ajoutées lors d'une restauration apparaissent dans le texte, mais, lorsque cela était possible, ces pages n'ont pas été filmées.

- [ ] Additional comments /
  Commentaires supplémentaires:

- [ ] Coloured pages / Pages de couleur

- [ ] Pages damaged / Pages endommagées

- [ ] Pages restored and/or laminated /
  Pages restaurées et/ou pelliculées

- [✓] Pages discoloured, stained or foxed /
  Pages décolorées, tachetées ou piquées

- [ ] Pages detached / Pages détachées

- [✓] Showthrough / Transparence

- [ ] Quality of print varies /
  Qualité inégale de l'impression

- [ ] Includes supplementary material /
  Comprend du matériel supplémentaire

- [ ] Pages wholly or partially obscured by errata slips, tissues, etc., have been refilmed to ensure the best possible image / Les pages totalement ou partiellement obscurcies par un feuillet d'errata, une pelure, etc., ont été filmées à nouveau de façon à obtenir la meilleure image possible.

- [ ] Opposing pages with varying colouration or discolourations are filmed twice to ensure the best possible image / Les pages s'opposant ayant des colorations variables ou des décolorations sont filmées deux fois afin d'obtenir la meilleure image possible.

This item is filmed at the reduction ratio checked below /
Ce document est filmé au taux de réduction indiqué ci-dessous.

10x			14x			18x			22x			26x			30x		
		12x			16x			20x			24x ✓			28x			32x	

1	2	3

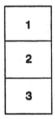

1	2	3
4	5	6

MICROCOPY RESOLUTION TEST CHART

(ANSI and ISO TEST CHART No. 2)

APPLIED IMAGE Inc

1653 East Main Street
Rochester, New York   14609   USA
(716) 482 - 0300 - Phone
(716) 288 - 5989 - Fax

CANADA
DEPARTMENT OF MINES
MINES BRANCH
Hon. W. Templeman, Minister; A. P. Low, LL.D., Deputy Minister; Eugene Haanel, Ph. D., Director.

BULLETIN No. 1.

INVESTIGATION OF THE PEAT BOGS AND PEAT INDUSTRY OF CANADA DURING THE SEASON 1908-9

SECOND EDITION

BY

ERIK NYSTRÖM, M.E.
AND
A. ANREP, PEAT EXPERT

OTTAWA
GOVERNMENT PRINTING BUREAU
1909

No. 30.

5 JUIN 1957

CANADA
EPARTMENT OF MINES
MINES BRANCH
Hon. W. Templeman, Minister; A. P. Low, LL.D., Deputy Minister;
Eugene Haanel, Ph.D., Director.

BULLETIN No. 1.

INVESTIGATION OF THE PEAT BOGS AND PEAT INDUSTRY OF CANADA, DURING THE SEASON 1908-9.

SECOND EDITION

BY

ERIK NYSTRÖM, M.E.,

AND

S. A. ANREP, PEAT EXPERT.

OTTAWA
GOVERNMENT PRINTING BUREAU
1909   No. 39.

LETTER OF TRANSMITTAL.

OTTAWA, March 8, 1909.

SIR,—

During the season of 1908, I, with Mr. A. Anrep as assistant, started the investigation of our peat resources. The bogs first to be considered were those which were favourably located as to transportation and market.

Several of these bogs are of considerable extent, and in order to arrive at a fairly accurate estimate as to their extent, depth and quality of the peat contained, a great number of holes have to be drilled.

The drilling, as well as the cutting out of the necessary lines, especially where the bogs are heavily wooded, requires considerable time, and these investigations will, therefore, involve many years' work.

Part of the summer was spent at Victoria Road peat bog, where the Anrep peat pulping machine imported from Sweden was set up, in order to manufacture sufficient peat fuel to start the Körting peat gas producer plant at present under erection at Ottawa.

During the summer I also visited the greater number of the peat plants so far erected in Canada, and in October attended the meeting of the American Peat Society held in Toledo, Ohio.

In the following report the bogs so far investigated are separately described, and accompanied by maps showing their areas, as well as the depths of the peat and the average degree of humification for each drill hole.

Yours respectfully,

(Signed) E. NYSTROM,

DR. EUGENE HAANEL,
Director of Mines,
Ottawa.

CONTENTS.

ILLUSTRATIONS.

MAPS.

INVESTIGATION
OF THE
PEAT BOGS AND PEAT FUEL INDUSTRY
OF CANADA, 1908.

ERIK NYSTRÖM and S. A. ANREP.

METHOD OF INVESTIGATION.

In order to map the surface of the bogs, lines were run at certain intervals, generally 1,000 feet apart. Drillings were made in each of these lines at intervals of 500 feet, and samples of the peat collected from different depths. Unless the peat was very different in appearance, all these samples were made into one general sample covering a certain section of the bog; but the degree of humification was noted for each drill hole, and for the different depths.

The different degrees of humification are expressed in letters in accordance with the following scale:—

C		B	
C+	indicating a	B +	indicating a peat
B C —	peat more or	A B —	more or less suitable
B C	less suitable	A B	for peat fuel.
B C +	for moss litter.	A B +	
B —		A —	
		A	

The peat classified in accordance with this scale from C to B— is only suitable for the manufacture of moss litter or similar products, and that from B to A for peat fuel. B, for instance, indicates that the peat fuel produced from such peat is light and consequently bulky, and A B a heavier peat, well suited for the manufacture of fuel. In the same manner C indicates a peat, or rather moss, well suited for moss-litter, and B— a less suitable material. The signs + and — after the letter respectively increase or decrease the degree indicated.

For the purpose in view, viz., to obtain a fairly accurate idea of our peat resources, it is quite sufficient, at least for the larger bogs, to run the lines 1,000 feet, or even further apart, as it would otherwise require a very long time to reach any conclusions; but where manufacture of peat products is intended, drillings ought to be made closer.

In Canada the method chiefly used for the manufacture of peat fuel has been to harrow the surface of the bog, or some similar method, and collect the partly air-dried peat in horizontal layers. In such cases the part of the bog which is intended to be worked should be carefully investigated, and the quality and properties of the different layers separately ascertained. The peat, in many cases, is of different nature in different layers, even in the same bog. and in several cases some layers are entirely unsuitable for the method above referred to. In Europe the method generally used for the production of peat fuel is different; the peat is there dug out of the bog in vertical trenches, the different layers of peat are mixed, and a uniform product is obtained.

CANADIAN BOGS.

The vegetation forming our bogs is in most cases identical, or very similar to that forming the European bogs.

In the following list, supplied by Mr. J. M. Macoun of the Geological Survey, the principal plants forming our bogs are enumerated:—

CHARACTERISTIC PLANTS OF THE PEAT BOGS OF ONTARIO.

While the number of species of Sphagnum growing on our eastern bogs is very large, the general bog covering is made up of a few species only, of which the following are the most abundant:—

Sphagnum acutifolium, Russ. and Warnst.
Sphagnum recurvum, Russ. and Warnst. and its varieties.
Sphagnum fuscum, von Klinggraeff, and its varieties.
Sphagnum cymbifolium, Ehrh.
Sphagnum medium, Limpr.

Next to Sphagnum the moss family which is most common is Hypnum. More than a score of species are to be found on almost any large bog, but many of these are not in sufficient abundance to affect the quality of the peat in any great degree. The most important are:—

Thuidium Blandowii, Bruch and Schimp.
Camptothecium nitens, Schimp.
Hypnum fluitans, Linn.

On tussocks and other raised parts of bogs a great variety of other mosses and lichens are always to be found, but few of these are worth noting, as they do not form large masses. Almost any hummock will be found to be composed chiefly of the following:—

Polytricum strictum, Banks, Hair-cap moss.
Cladonia rangiferina, Hoffm., Reindeer moss.

Other species of Polytricum and Cladonia will almost always be noted, but those mentioned form the mass.

The order Cyperaceæ is always well represented on and around bogs. In the wetter parts, where there is no Sphagnum covering, many species of Carex are always to be found, among them: C. utriculata, Boott; C. vulpinoidea, Michx.; C. teretiuscula, Good; C. Sartwellii, Dewey; C. riparia, Curtis; and C. aquatilis, Wahl.

On the bog itself the most important species are:—

Carex exilis, Wahl.
Carex sterilis, Willd.
Carex oligosperma, Michx.
Carex trisperma, Dewey.

The only other genus of Cyperaceæ that is represented is Eriophorum. of which four or five species occur on every bog. Of these the most valuable is:—

Eriophorum callitrix, Cham.

or E. vaginatum, as it was formerly called. The other species are:—

Eriophorum Chamissonis, C. A. Meyer.
Eriophorum gracile, Roth.
Eriophorum angustifolium, Roth.
Eriophorum Virginicum, L.

While not properly bog plants, a large number of aquatic species[1] grow in ponds, or wet, marshy spots in the bog or along its borders. These are frequently mixed together, but often large areas will be covered by one or two species, such as: Typha, Sagittaria, Sparganium, Utricularia, or Potamogeton.

Only the most abundant species are enumerated below:—

Sagittaria latifolia, Willd . . . . Broad-leaved arrowhead.
Utricularia vulgaris, L . . . . Greater bladderwort.
Iris versicolor, L . . . . Larger blue flag.
Scirpus lacustris, L . . . . Lake bulrush.
Sparganium simplex, Michx . . . . . Bur-reed.
Typha latifolia, L . . . . Cat-tail.
Glyceria nervata, Willd . . . . Nerved manna-grass.
Glyceria Americana, Torr . . . . Tall manna-grass.
Potamogeton, Several species . . . Pond-weed.
Scirpus cyperinus, Kunth . . . . Wool-grass.
Nymphaea odorata, Ait . . . . White water lily.
Nuphar advena, Ait . . . . Yellow water lily.

Of herbaceous plants there are more than 100 species, in addition to those mentioned above, which may properly be called bog plants, but the majority of these are either so insignificant in size or grow so far apart that they effect no appreciable change in the composition of the peat. The most common and valuable species are:—

Menyanthes trifoliata, L . . . . Buckbean.
Drosera rotundifolia, L . . . . Round-leaved sundew.
Chiogenes hispidula, T. and G . . . Creeping snowberry.
Scheuchzeria palustris, L . . . . Scheuchzeria.
Habenaria dilatata, Pursh . . . . . . Tall white bog orchis.
Habenaria hyperborea, L . . . . . . Tall green orchis.

[1] In cases where the bogs have been formed by the growing over of lakes the remains of these plants generally form the deeper layers of the bogs.

Vaccinium macrocarpon, Ait. . . . Large cranberry.
Vaccinium oxycoccus, L . . . . Small cranberry.
Sarracenia purpurea, L . . . . Pitcher plant.

Even the most open bogs are usually covered in great part by shrubs or shrubby plants, and in boggy woods they occur frequently in even greater profusion. The most widely distributed and best known species are:—

Ledum Groenlandicum, O. Eder. . Labrador tea.
Andromeda glaucophylla, Link. . Andromeda.
Kalmia glauca, L . . . . Mountain laurel.
Cassandra calyculata, L . . . . Leather leaf.
Spiræa salicifolia, L . . . . Meadowsweet.
Vaccinium Canadense, Rich. . . . . Canada blueberry.
Vaccinium Pennsylvanicum, Lam. Low blueberry.
Vaccinium nigrum, Britton . . . . . Low black blueberry.
Salix myrtilloides, L. . . . . Myrtle-leaved willow.

The only trees that may be said to be truly characteristic of bogs are black spruce, Picea Mariana, B.S.P. and tamarack, Larix laricina, Koch., but along the margins of most bogs, and on many old bogs where there is good drainage, cedar, hemlock, balsam, ash, and a few other species will generally be found.

DESCRIPTION OF INDIVIDUAL PEAT BOGS.

MER BLEUE PEAT BOG.

This bog is situated about 8 miles from Ottawa, Ont., in the townships of Gloucester and Cumberland (see accompanying map) and covers more or less of:—

Lots 13-16, con. III, township of Gloucester.
Lots   3-15, con. IV, " "
Lots   1-16, con. V, "
Lots   1- 9, con. VI. " "
Lots   1- 2, con. VII, " "
Lots   9-16, con. XI, township of Cumberland.
Lots 12-16, con. X. " "

The total area covered by this bog is, approximately, 5,004 acres. Of this area

1,564 acres have a depth of less than 5 feet; average depth 2'-9"
2,237 acres have a depth of 5 to 10 feet, average depth 6'-8"
 856 acres have a depth of 10 to 15 feet, average depth 11'-8"
 347 acres have a depth of more than 15 feet, average depth 16 feet.

The volume of the peat contained is:—

In the area with a depth of less than 5 feet . . . .   6,938,946 cub. yds.
5 to 10 feet deep. . . . . 24,036,117 " "
10 to 15 feet deep. . . . . 16,102,611 " "
more than 15 feet deep . . . .   8,973,037 " "

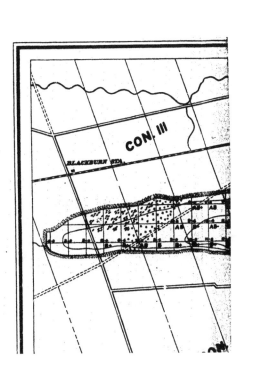

CANADA
DEPARTMENT OF MINES

MER BLEUE PEAT BOG

CANADA
TMENT OF MINES
MINES BRANCH
INISTER; A. P. Low, LL.D. DEPUTY MINISTER;
E. HAANEL, PH.D., DIRECTOR.

JE PEAT BOG

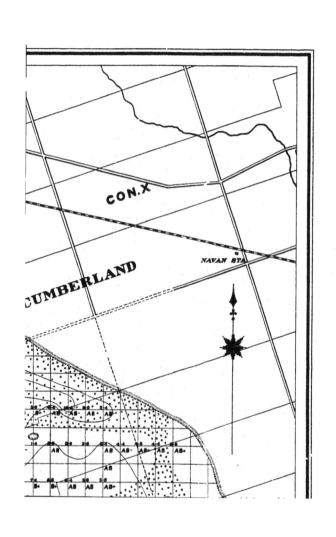

CON. X

NAVAN STA.

CUMBERLAND

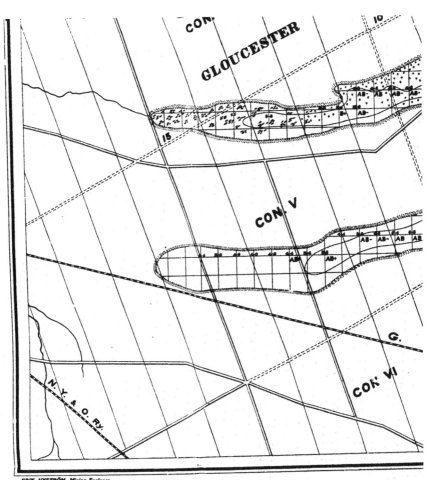

ERIK NYSTRÖM, Mining Engineer.
S. A. ANREP, Peat Expert.

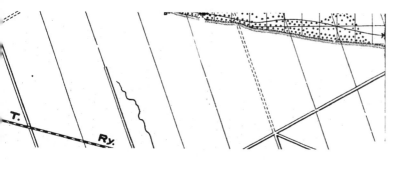

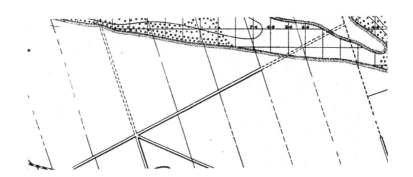

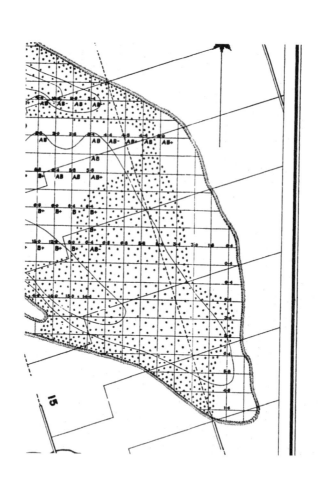

The peat is fairly well humified and uniform in quality, and, with proper treatment, will produce a comparatively good peat fuel.

The peat, after the bog is thoroughly drained, will probably settle down from one to three feet.

Deducting, therefore, the 1,564 acres, which at present have an average depth of about three feet and which are not likely to be profitably worked by machinery, and also allowing for the decrease in depth through drainage, we have left:—

> 2,237 acres with an average depth of approximately . . . 5 feet.
> 856 acres with an average depth of approximately . . .9½ "
> 347 acres with an average depth of approximately . . .13 "

with a total volume of 38,442,494 cubic yards of peat.

Assuming that one cubic yard of such drained bog will furnish 200 lbs. dry peat substance, the total tonnage of dry peat substance available is 3,844,249 tons of 2,000 lbs., or 5,125.655 tons peat fuel, with 25% moisture.

The bog consists of three distinct basins, partly separated by two comparatively high ridges. These ridges are east of lot 5 in Gloucester township, partly submerged in the bog, but in places form what may properly be called islands in the bog. The bottom of the bog is a compact blue clay.

The principal vegetation forming the peat in this bog is Sphagnum, more or less mixed with Eriophorum and most of the other plants given in the preceding tables.

Large areas of the surface of the bog are covered by a more or less heavy growth of spruce and tamarack, and the peat itself is intermixed with large quantities of roots, trunks, and stumps of trees.

Analyses of Peat (absolutely dry).

	1	2	3	4	5	6	7
Volatile matter	65·90	67·57	68·40	63·22	68·78	68·73	69·49
Fixed carbon	24·22	25·35	25·00	24·86	25·73	26·27	26·04
Ash	10·88	7·18	6·60	11·92	5·51	5·00	4·47
Phosphorus		0·026				0·024	
Sulphur		0·314				0·317	
Nitrogen		1·40				1·13	
Calorific value B.T.U. per lb.	8821	9021		8805	9126	9444	9301

The content of ash, although in some cases comparatively high, is not excessive, and the calorific value is satisfactory.

In order to work this bog profitably on a larger scale, it has to be thoroughly drained. Such drainage will, undoubtedly, involve a large expenditure of money, but considering the value of the land that would be recovered and which at present is practically valueless, and also the improvement which would result in the surrounding farming land through such drainage, such an undertaking would eventually be a paying proposition. The greater part of the bog could then be utilized, either for the manufacture of peat fuel, for agricultural purposes, or both.

The bog is exceedingly well situated both in regard to shipping facilities and probable market, the centre of the bog being only about eight miles from Ottawa. The Canadian Pacific railway (Ottawa-Montreal line) passes on the north side of the bog, and the Grand Trunk railway (Ottawa-Montreal line) on the south side.

ALFRED PEAT BOG.

This bog is situated about 40 miles from Ottawa, in the townships of Alfred and Caledonia (see accompanying map) and covers more or less of:—

Lots 6-10, con. VII, township of Alfred.
Lots 6-13, con. VIII, "   "
Lots 8-13, con. IX, "   "
Lots 13-24, con. I, township of Caledonia.
Lots 9-24, con. II, "   "
Lots 9-18, con. III, "   "

also extending into the township of Longueuil.

The total area covered by this bog is approximately 6,800 acres. Of this area

1,377 acres have a depth of less than 5 feet, average depth 3'-10"
3,084 acres have a depth of 5 to 10 feet, average depth 8 feet.
1,316 acres have a depth of 10 to 15 feet, average depth 11'-10"
1,014 acres have a depth of more than 15 feet, average depth 16'-5"

The volume of the peat contained is:—

In the area with a depth of less than 5 feet .... 8,441,928 cub. yds.
5 to 10 feet deep .... 39,800,592 "   "
10 to 15 feet deep .... 25,093,136 "   "
more than 15 feet deep .... 26,846,800 "   "

The peat in the part of the bog located in Alfred township is, with the exception of that immediately surrounding the small pond shown on the map, comparatively well humified and will produce a good fuel.

The upper layers of the part of the bog located in Caledonia township are, especially in the centre of that part, comparatively poorly humified, and a fairly light fuel can there be expected.

Deducting the 1,377 acres with a depth of less than five feet, and allowing for the decrease in depth through drainage, we have left:—

3,084 acres with an average depth of approximately ... 6 feet.
1,316 acres with an average depth of approximately ... 9 "
1,014 acres with an average depth of approximately ...13 "

with a total volume of 70,270,200 cubic yards of peat.

Calculating that one cubic yard of such drained bog will furnish 200 lbs. dry peat substance, the total tonnage of dry peat substance available is 7,027,000 tons of 2,000 lbs., or 9,369,000 tons peat fuel, with 25% moisture.

The bog consists of one large basin, the deeper part of which is located in Caledonia township. In the northern part of the bog in Alfred township a

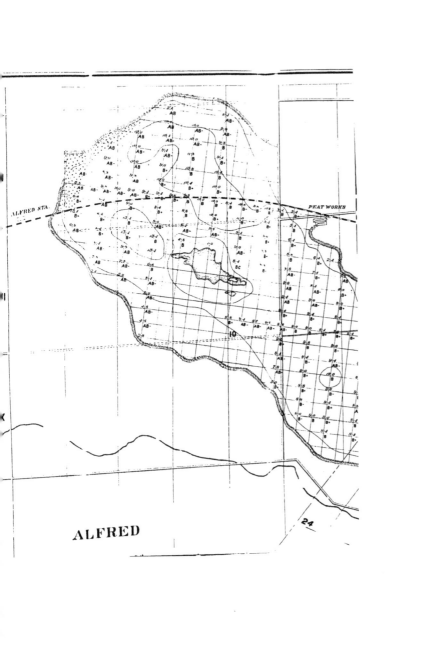

ALFRED

24

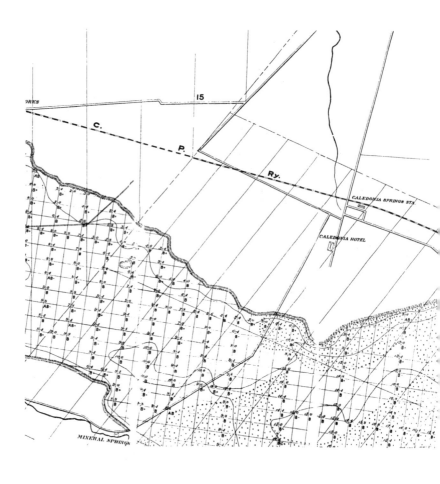

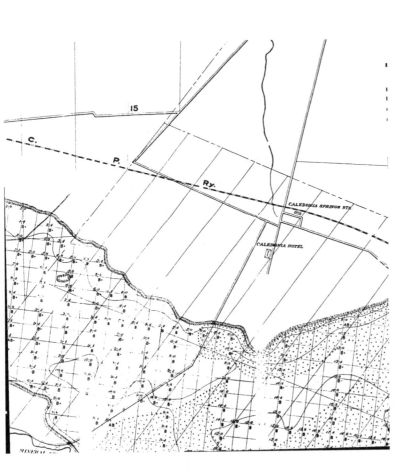

LONGEUIL

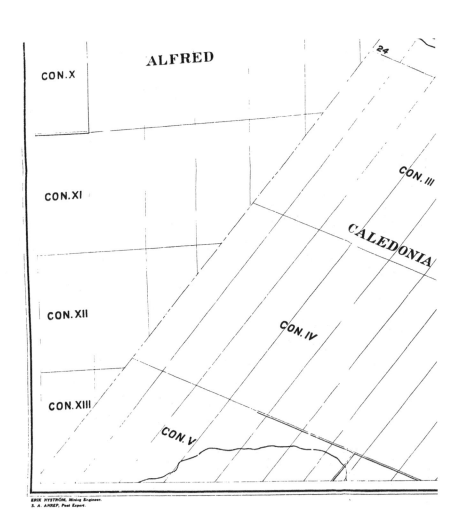

ALFRED

CON. X

CON. XI

CON. XII

CON. XIII

CON. III

CALEDONIA

CON. IV

CON. V

24

ERIK NYSTRÖM, Mining Engineer.
S. A. ANREP, Peat Expert.

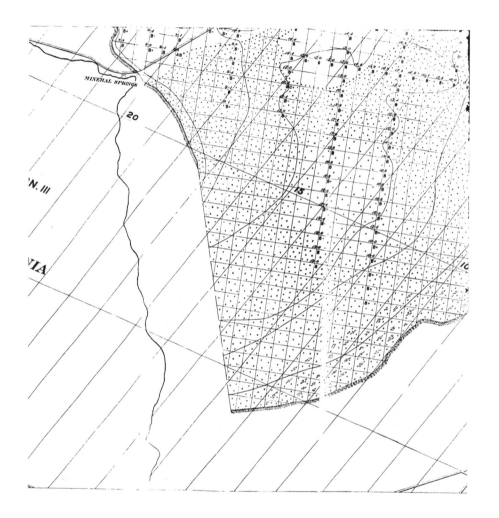

MINERAL SPRINGS

20

15

Scale 5 inches 1 mile

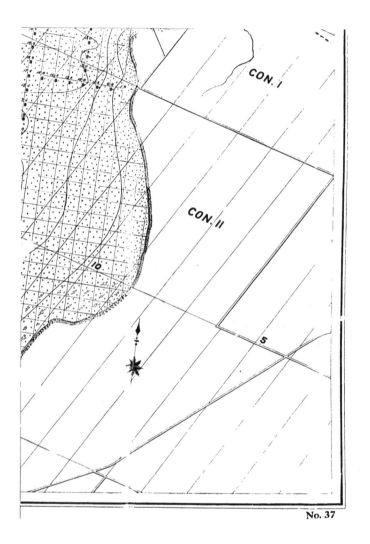

CON. I

CON. II

10

5

No. 37

comparatively deep depression also occurs, and the peat contained in this part of the bog is well suited for the production of fuel.

The peat is principally formed by Sphagnum, but in certain parts of the bog it is also mixed with Eriophorum, Hypnum, Carex, and, especially around the small pond, with the remains of typical aquatic plants. The northwestern part of the bog in Alfred township, for instance, is strongly mixed with Eriophorum, Hypnum, and, towards the margin, with Carex. The southeastern part of the bog in the same township is principally formed by Sphagnum, only slightly mixed with Eriophorum and the remains of grasses. The northern part of the bog in Caledonia township is strongly mixed with Hypnum, and the southern and eastern part with the remains of plants, such as Vaccinium and similar plants. In Caledonia township the bog is more or less heavily wooded with spruce and tamarack, and towards the margin with cedar and hardwood trees.

Certain parts of the bog contain large amounts of roots and stumps, especially in the southern part of Caledonia township, where in places it was impossible to get the drill through to the bottom, on account of these obstacles.

The bottom of the bog is a compact blue clay.

Analyses of Peat (absolutely dry).

	1	2
Volatile matter	68·13	68·72
Fixed carbon	26·56	24·22
Ash	5·31	7·06
Phosphorus	0·029	0·022
Sulphur	0·292	0·375
Nitrogen	1·23	1·92
Calorific value B.T.U. per lb	8730	9058

The content of ash is about the average, and the calorific value satisfactory.

Part of this bog in Alfred township is traversed by the Canadian Pacific railway (Ottawa-Montreal line), which also touches a part of the bog in Caledonia township. As regards transportation facilities the bog is, therefore, well situated, and the distance to Ottawa being only some forty miles the freight rates ought not to be excessive.

The greater portion of the bog in Alfred township is already fairly well drained and the balance of the water could probably be handled by a pump. Large areas of the shallow parts of the bog are yearly burnt off by the farmers and the land utilized for agricultural purposes.

Some years ago the Montreal and Ottawa Peat Company erected a peat plant at this bog, and started to manufacture peat briquettes in accordance with the Dobson method. (For description of this method see Bulletin No. 5 published by the Ontario Bureau of Mines, Toronto, and also Report on Peat and Lignite, by E. Nystrom, published by the Mines Branch, Department of

14

Mines, Ottawa). Only a few hundred tons of briquettes were made, however, and the plant has been closed down for several years.

The Company owns the north half of lots 9, 10 and 11, con. VIII, township of Alfred.

The bog contains a large amount of the fibrous and less humified variety of peat, which is less suitable for the olson process.

WELLAND PEAT BOG.

This bog is situated in the townships of Wainfleet and Humberstone, six miles north of the town of Welland, and between the Welland canal and its feeder (see accompanying map) and covers more or less of:—

Lots 1- 8, con. IV, township of Wainfleet.
Lots 1-13, con. III,   "   "
Lots 4-13, con. II,   "   "
Lots 27-33, con. IV, township of Humberstone.
Lots 27-33, con. III,   "   "

The total area covered by this bog is approximately 4,900 acres. Of this area

1,423 acres have a depth of less than 5 feet, average depth 3 feet.
2,877 acres have a depth of 5 to 10 feet, average depth 7'–3"
588 acres have a depth of more than 10 feet, average depth 11 feet.

The volume of the peat contained is:—

In the area with a depth of less than 5 feet ... 6,885,170 cub. yds.
5 to 10 feet deep......................... 33,662,231 "   "
more than 10 feet deep..................... 10,427,600 "   "

In the western and eastern parts of the bog and around the margins the peat is well humified, and is mainly formed by the remains of Hypnum and Eriophorum, with only a small amount of Sphagnum, Carex and aquatic plants. In the central part of the bog the peat is mainly formed by Sphagnum, with small amounts of Eriophorum and other plants. The peat in this part, especially in the upper layers, is very poorly humified and unsuitable for the manufacture of fuel, but will furnish a fairly good material for the production of moss litter.

The bog, as shown above, is comparatively shallow, and deducting the 1,423 acres with a depth of less than 5 feet, allowing for the decrease in depth through drainage, and assuming that the balance could be utilized, we have left:—

2,877 acres with an average depth of approximately ....5 feet.
588 acres with an average depth of approximately ....8 "
with a total volume of 30,796,480 cubic yards of peat.

With the assumption that one cubic yard of such drained bog will furnish 200 lbs. dry peat substance, the total tonnage of dry peat substance is 3,079,600 tons of 2,000 lbs., or 4,106,000 tons peat fuel, with 25% moisture.

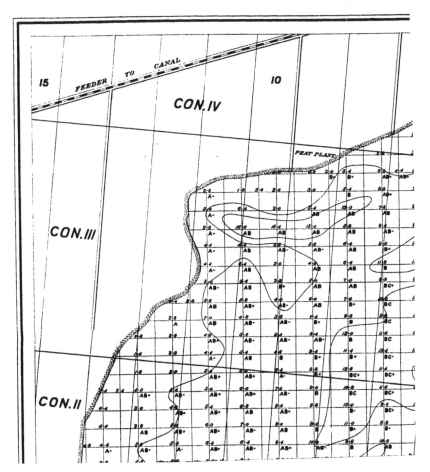

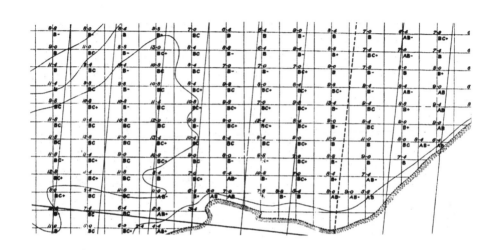

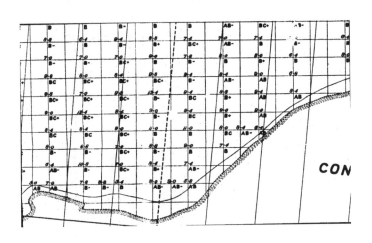

CON

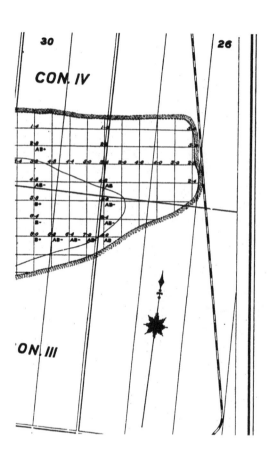

CON.II

10

ERIK NYSTRÖM, Mining Engineer.
S. A. ANREP, Peat Expert.

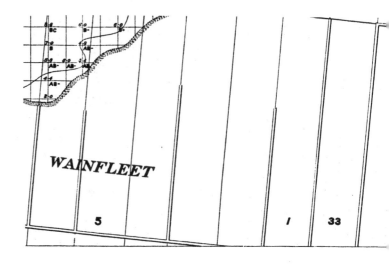

WAINFLEET

5   /   33

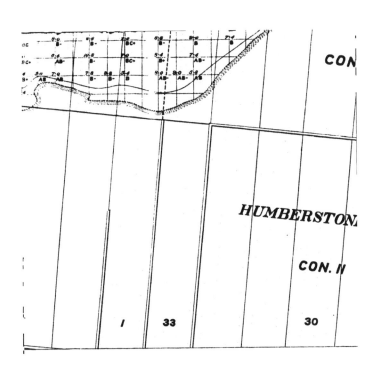

CON

HUMBERSTON

CON. I

1   33   30

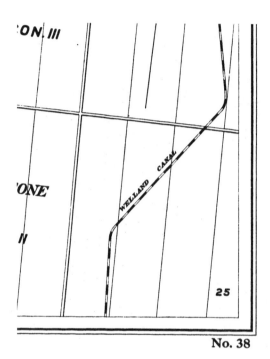

:ON. III

'ONE

II

WELLAND CANAL

25

No. 38

A very great amount of roots and stumps of trees are intermixed with the peat and large areas have been burnt to considerable depth. Taking into account also the large area of poorly decomposed peat, this bog is not well adapted for the manufacture of fuel on a large scale.

The bog consists of one large basin with clay bottom. The surface of the bog is comparatively free from growing trees with the exception of some small areas in its central parts and around the margin.

Analyses of Peat (absolutely dry).

	1	2	3
Volatile matter	67·14	70·90	70·53
Fixed carbon	26·48	24·84	24·28
Ash	6·38	4·26	5·19
Phosphorus	0·027	0·024	
Sulphur	0·317	0·248	
Nitrogen	1·13	1·74	
Calorific value B.T.U. per lb.	9118	8596	8667

The bog is held under lease by the Peat Industries, Ltd., of Brantford, which Company some years ago erected a plant and carried out extensive experiments. The results obtained must have been unsatisfactory, since no work has been done for several years. (For description of the methods used, see Bulletin No. 5, published by the Ontario Bureau of Mines, and Report on Peat and Lignite by E. Nystrom).

NEWINGTON PEAT BOG.

This bog is situated in the townships of Osnabruck, Roxborough, and Cornwall, county of Stormont, on the New York and Ottawa railway, some 40 miles from Ottawa, close to the village of Newington (see accompanying map) and covers more on less of

Lots 1– 5, con. VII, township of Osnabruck.
Lots 1– 6, con. VII, " "
Lots 2– 3, con. VI, " "
Lots 17–35, con. II, township of Roxborough.
Lots 17–39, con. I, " "
Lots 20–23, con. IX, township of Cornwall.
Lots 26–30, con. IX, " "
Lots 31–36, con. VI, "

The total area covered by this bog is approximately 3,800 acres. Of this area

929 acres have a depth of less than 5 feet, average depth 2 feet.
1,191 acres have a depth of 5 to 10 feet, average depth 8 feet.
748 acres have a depth of 10 to 15 feet, average depth 12 feet.
696 acres have a depth of 15 to 20 feet, average depth 17'-4"

158 acres have a depth of 20 to 25 feet, average depth 21' 4"

120 acres have a depth of more than 25 feet, average depth 26 feet.

The volume of the peat contained is:—

In the area with a depth of less than 5 feet. .   2,996,704 cub. yds.

5 to 10 feet deep.   15,378,518   "   "

10 to 15 feet deep.   14,491,111   "   "

15 to 20 feet deep.   19,542,777   "   "

20 to 25 feet deep.   5,464,926   "   "

more than 25 feet deep.   5,037,777   "   "

The western part of the bog in Osnabruck township is a pure Sphagnum bog. The northern part in the same township is badly damaged on the surface by fires, and more or less unsuitable for the manufacture of fuel by ordinary processes, at least where the surface of the bog is used for a drying field.

The western part of the bog in Roxborough township is principally formed by Sphagnum, only slightly mixed with the remains of aquatic plants. This part will furnish fairly good material for the manufacture of moss-litter. The eastern part of the bog in the same township is heavily wooded and of a swampy character. The peat is here mixed with herbaceous plants.

The bog is fairly free from stumps and roots, and large areas are, as shown above, of considerable depth. With the exception of the part of the bog in the western part of Roxborough township, the peat is fairly well humified and suitable for fuel.

Deducting the 929 acres with a depth of less than 5 feet, and allowing for the decrease in depth through drainage, we have left:—

1,191 acres with an average depth of approximately . . . . . 6 feet.

748 acres with an average depth of approximately 9   "

696 acres with an average depth of approximately 14   "

158 acres with an average depth of approximately 17   "

120 acres with an average depth of approximately 21   "

with a total volume of 46,566,478 cubic yards of peat.

With the assumption that one cubic yard of such drained bog will furnish 200 lbs. dry peat substance, the total tonnage of dry peat substance is 4,656,000 tons of 2,000 lbs., or 6,208,800 tons peat fuel, with 25% moisture.

Analyses of Peat (absolutely dry).

	1	2	3	4	5	6	7
Volatile matter	66·75	67·07	68·84	71·42	69·54	65·77	66·97
Fixed carbon	25·77	26·27	26·65	24·14	26·75	27·30	26·70
Ash	7·48	6·66	4·51	4·24	3·71	6·93	6·34
Phosphorus	0·028	0·030		0·032			
Sulphur	0·530	0·494		0·348			
Nitrogen	1·85	1·80		1·63			
Calorific value B.T.U. per lb.	8721	8165	8877	8636	9102	8240	8312

MICROCOPY RESOLUTION TEST CHART

(ANSI and ISO TEST CHART No. 2)

1653 East Main Street
Rochester, New York   14609   USA
(716) 482 - 0300 - Phone
(716) 288 - 5989 - Fax

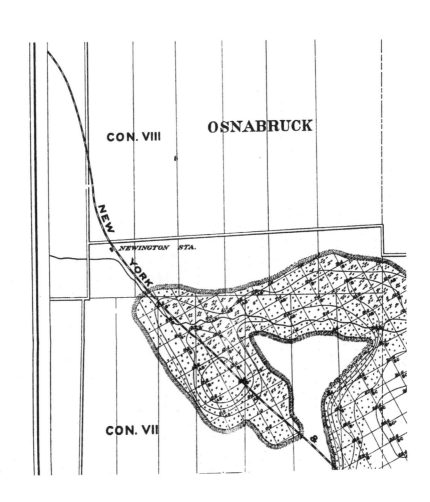

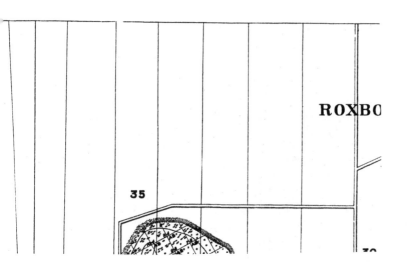

ROXBO

35

ROXBOROUGH

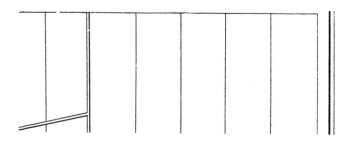

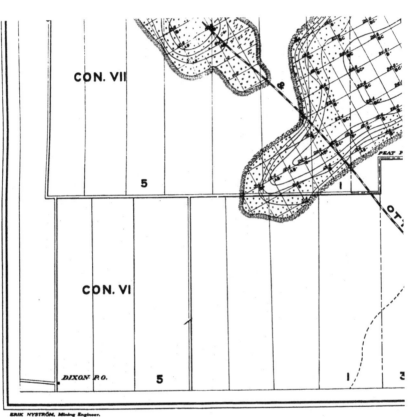

CON. VII

5

CON. VI

DIXON P.O.   5   I

ERIK NYSTRÖM, Mining Engineer.
S. A. ANREP, Peat Expert.

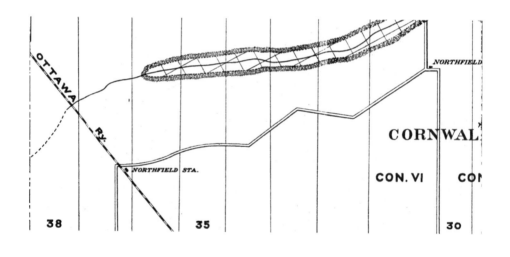

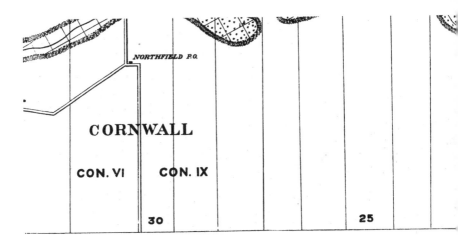

NORTHFIELD P.O.

CORNWALL

CON. VI   CON. IX

30   25

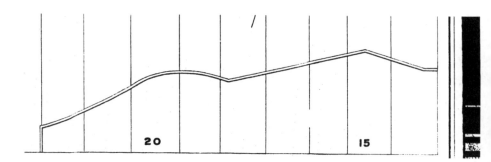

17

A peat plant was erected at this bog (see map) by the ominion 'eat Products Company, Ltd., of Brantford. A description of the method intended to be used is given in the Bulletin on Peat issued in 1904 by the Geological Survey. The peat was to be cut out of the bog by means of Dolberg cutting machines, and conveyed, by means of a conveyer, to the plant, where it was put through a Dolberg peat machine. The peat bricks were then loaded on iron trucks, which were made to fit a tunnel furnace, where the drying was to be accomplished. In order to dry the peat, hot air was forced through the furnace by means of a fan. The impossibility of such a method is evident from the calculation given on page 24. The process was claimed to be in successful operation in Europe, but although processes of this kind have been tried, none have proved successful.

For several years the plant has been closed down.

PERTH PEAT BOG.

This bog is situated in the township of rummond, about a mile and a half north of the town of Perth, and covers (see accompanying map) more or less of:—

Lots 1– 5, con. III, township of rummond.
Lots 2– 6, con. IV, " "
Lots 3–11, con. V, " "
Lots 4–13, con. VI, " "
Lots 12–13, con. VII, " "

The total area covered by this bog is approximately 3,800 acres. Of this area

678 acres have a depth of less than 5 feet, average depth 2'–8".
958 acres have a depth of 5 to 10 feet, average depth 8 feet.
2,098 acres have a depth of 10 to 15 feet, average depth 11 feet.
106 acres have a depth of more than 15 feet, average depth 16 feet.

The volume of the peat contained is:—

In the area with a depth of less than 5 feet   3,064,133 cub. yds.
5 to 10 feet deep .... 12,362,963 " "
10 to 15 feet deep .... 37,235,000 " "
more than 15 feet deep .... 2,860,888 " "

The peat is formed by Sphagnum, Hypnum and Erio horum, in places mixed with the remains of aquatic plants and grasses. T peat in the open part of the bog is principally formed by Sphagnum and Hypnum, only slightly mixed with Eriophorum, but in the wooded parts of the bog the peat is more mixed with the remains of various plants and also with roots and trunks of trees. The surface of the latter parts is covered with a layer of Sphagnum moss, 1 to 2 feet in thickness. Certain parts of the bog, which are more or less drained, are heavily grown over with willow brushes, and in places with hardwood trees. The peat suitable for the manufacture of fuel is comparatively little humified, and the fuel produced will be comparatively light.

2

educting the 678 acres with a depth of less than 5 feet, and allowing for the decrease in depth through drainage, we have left:—

958 acres with an average depth of approximately . . . . . 6 feet.
2,098 acres with an average depth of approximately . . . . . 8 "
106 acres with an average depth of approximately . . . . .12 "
with a total volume of 38,445,222 cubic yards.

With the assumption that one cubic yard of such drained bog will furnish 200 lbs. dry peat substance, the total tonnage of dry peat substance is 3,844,500 tons of 2,000 lbs., or 5,126,000 tons peat fuel, with 25% moisture.

Analyses of Peat (absolutely dry).

	1	2
Volatile matter	70·34	71·51
Fixed carbon	25·35	24·60
Ash	4·31	3·89
Phosphorus	0·030	0·027
Sulphur	0·405	0·334
Nitrogen	1·66	1·94
Calorific value B.T.U. per lb.	9067	9148

A peat plant was erected at this bog (see map) by the Lanark County Peat Fuel Company of Perth. The part of the bog in the immediate vicinity of the plant had formerly been used for agricultural purposes and was consequently well drained. The peat had, however, been exposed to frost and its cohesive properties were very poor, and content of ash high.

Analysis of Peat from the Bog in the Vicinity of the Peat Plant.

Volatile matter . . . . .64·80%
Fixed carbon . . . . .21·74%
Ash . . . . .13·46%
Calorific value . . . . 8319 B.T.U. per lb.

To judge from the work done, hardly any peat fuel has been manufactured and the plant has been closed down for several years.

VICTORIA ROAD PEAT BOG.

This bog is situated in the townships of Bexley and Carden, about one mile from Victoria Road station on the Midland division of the Grand Trunk railway, and covers (see accompanying map) more or less of:—

Lots 3-4, con. I, township of Bexley.
Lots 3-4, con. X, township of Carden.

The total area covered by this bog is only 67 acres. Of this area 36 acres have a depth of less than 5 feet, average depth 2 feet.
15 acres have a depth of 5 to 10 feet, average depth 6'-4"

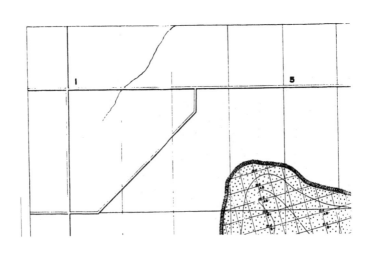

ANADA
NT OF MINES
ES BRANCH
; A. P. Low, Ll. D., Deputy Minister;
sri, Ph. D., Director

PEAT BOG

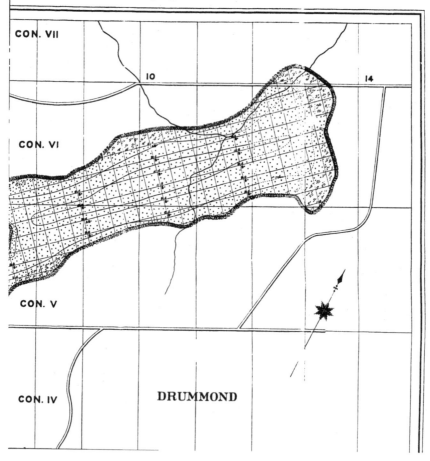

CON. VII

10

14

CON. VI

CON. V

CON. IV

DRUMMOND

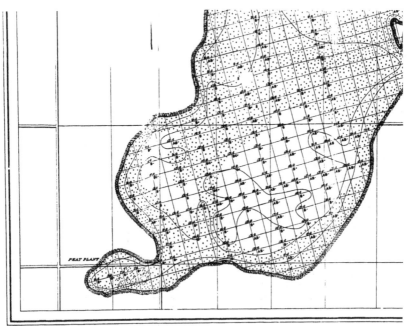

ERIK NYSTRÖM, Mining Engineer.
S. A. ANREP, Peat Expert.

PEAT PLANT

Scale 5 i

▦ Marg

▒ Heavi

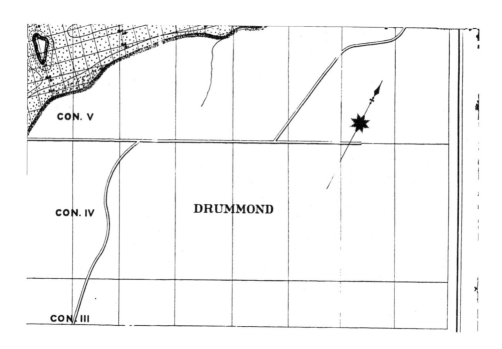

CON. V

CON. IV   DRUMMOND

CON. III

condensed fuel in the shape of briquettes, and for this purpose large amounts of m······ have been spent in experiments, but without reaching satisfactory ···

1···est methods so far tried for the above purpose are undoubtedly ···· ··· ··duced by Mr. A. Dobson, of Beaverton, and later, with some modifications, by Messrs. Milne and Dr. McWilliam, at Dorchester, near London, Ont.

Mr. Dobson's method and machinery are described in detail in Bulletin No. 5 of the Bureau of Mines of Ontario, and in Report on Peat and Lignite, by E. Nyström. At the present time, the plant at Beaverton, Ont., where Mr. ol·son first started, is the only one of his plants which is in operation, and that only occasionally. ····uring the last seasons, however, no peat fuel has been produced, but Mr. Dobson states that at Beaverton, and also at Great Meadows, N.J., peat mull is produced by his machinery, which product is used as a filler in the manufacture of fertilizers.

The Dorchester plant is situated about two miles from Dorchester station, near London, Ont., and is operated by Dr. J. McWilliam. This was the only plant in actual operation during part of last season. The method employed is briefly as follows:—

The surface of the bog is harrowed by means of a common harrow drawn by a horse. The peat is then exposed to the air and sun and left until partly air-dried. When sufficiently dried, the upper layer is collected by means of a suction fan mounted on an electrically drawn car moving on rails laid down on the bog, and deposited in a special car moving on the same rails.

The collector travels back and forth and collects the peat from the area which can be reached by the suction pipe of the fan. This pipe is connected with the fan by means of a flexible joint, and can be swung out at a greater or less distance from the track on which the car travels, thereby covering a considerable area from one track. The loaded car is brought to the plant, where the peat is deposited in a storage shed. From the shed it is conveyed to a drier and subsequently to the briquetting press.

The moisture in the peat collected from the bog is not, as a rule, uniform, and with the dryer used it was not easy to regulate the percentage of moisture in the peat going to the press, which, to some extent, may explain the irregular quality of the briquettes produced. A large percentage of the briquettes made are very solid and would probably stand shipping and rehandling, but others easily fall to pieces and produce a large amount of fine dust, which is objectionable, especially for domestic purposes. This large amount of dust produced is the principal defect with all the peat-briquettes so far produced, and up to the present time this difficulty has not been overcome.

The power for the plant is supplied by a 100 h.p. boiler, furnishing the steam for a 75 h.p. engine, which also drives the dynamo.

The machinery employed is gradually being improved, but it can hardly be claimed that this process is yet fully worked out.

PEAT PLANT

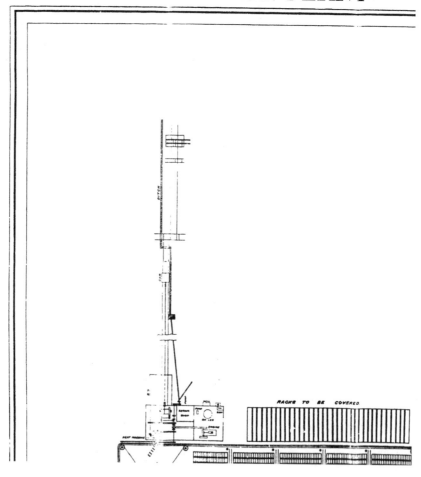

RACKS TO BE COVERED

CARDEN

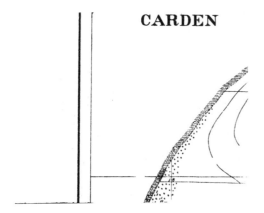

CARDEN

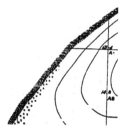

PEAT BOG

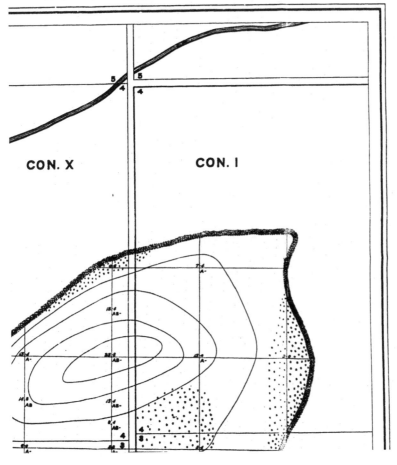

CON. X   CON. I

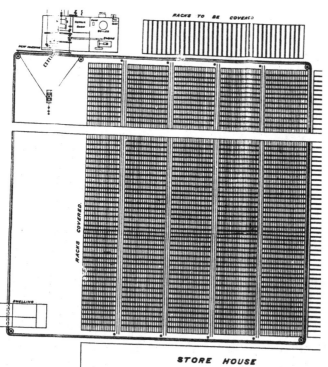

RACKS TO BE COVERED

RACKS COVERED.

DWELLING

STORE HOUSE

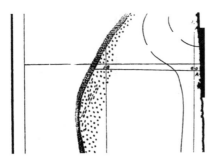

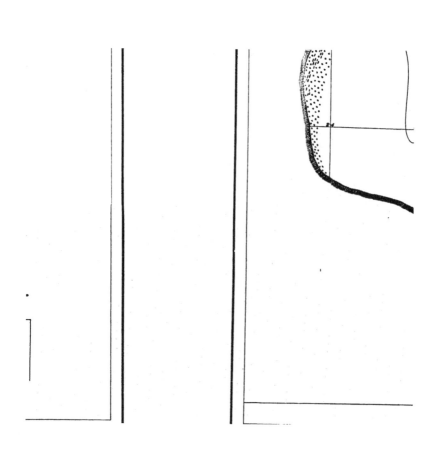

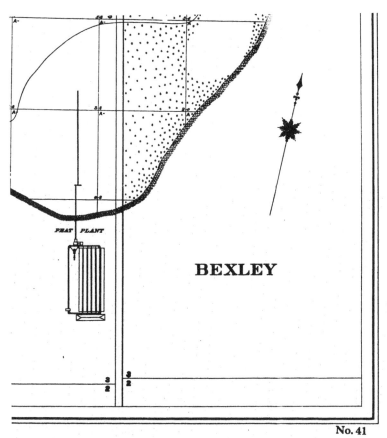

PEAT PLANT

BEXLEY

No. 41

Scale, 25 inches = 1 mile.

of the drying area, and cross tracks at the top or farther end, and at the bottom or nearer end, are provided, and there is also a means of connecting any up and down track with the end tracks, using specially designed portable corners in such a way as to form with the return track a rectangular course with sharply rounded corners, the length of which is the length of the drying area, and the width the distance from the return track to that particular up and down track in use.

This plant being the only one where power and drying arrangements could be obtained for the manufacture of air-dried peat, it was decided to install the Anrep peat machine imported by the epartment, as the plant was represented to be in good condition, and with a slight expenditure could be put in running order.

It was very soon found, however, that the condition of the plant was anything but good, the tracks, drying racks, and most of the other apparatus were very poorly built, and in order to make a fair trial of the plant more time and money than was available would have been required, the object being to manufacture sufficient peat fuel for the gas producer plant under erection and to test the Anrep peat machine. Work was, however, started in the early part of August. The peat machine fulfilled every expectation, and with proper arrangements could easily treat an amount of wet peat equivalent to 25 to 35 tons air-dried peat per day.

The principal objection to the plant as constructed by Mr. Moore is its complicated design. Even if the plant had been properly built, its practicability for the purpose is questionable. With sufficient area for drying purposes, the conveyer used is bound to cause trouble, and the unloading of the pallets from the cars on to the drying racks, as well as the subsequent transportation of the dried peat to the storehouse, is neither practical nor labour-saving. The excavator formerly used at this plant was dismantled and its operation and practical use could not be ascertained, but even assuming that mechanically it worked satisfactorily, it is questionable if the air-drying down to some 50 or 60% moisture before the peat is put through the pulping machine is advantageous. Such drying is liable to decrease the cohesive properties of the peat and cause it to crack and crumble to pieces more than would otherwise be the case.

GUELPH PEAT PLANT.

This plant is owned by the Imperial Peat Company, Ltd., and operations have recently been started. The object is to manufacture peat briquettes. No data as to the success of this Company are as yet to be had.

FARNHAM PEAT PLANT.

The plant near Farnham, Que., was erected by the United States Peat Fuel Company, with head office in Chicago. At my visit to this plant, during the summer, I was refused admittance, but it is evident that this Company, up to the present time at least, has no definite plan for the carrying out of the work.

LAC DU BONNET PEAT PLANT.

A peat plant, owned by the Inter-west Peat Fuel Co., of Winnipeg, Man., was erected at this place, but so far no practical results have been obtained. The plant is located at the Julius muskeg, east of Winnipeg, which covers an enormous area, but from information received it is probable that large areas of this bog are shallow and that the peat is more or less poorly humified.

DRYING, CARBONIZING, ETC.

The idea of the possibility of drying the wet peat as dug out of the bog, by means of artificial heat alone, seems still to prevail amongst a number of people interested in the peat industry. A very simple calculation shows, however, the impracticability of such an undertaking. Assuming that a drained bog contains $12\frac{1}{2}\%$ dry peat substance, which is a good average, 100 lbs. of wet peat contains $87\frac{1}{2}$ lbs. of water. Assuming further that 80% of the fuel value of the fuel used could be utilized, that 1,100 B.T.U. are required to evaporate 1 lb. water, and that the dry peat has a calorific value of 9,000 B.T.U. per lb.; in order to evaporate the water, we consequently require:—

$$\frac{87 \cdot 5 \times 1,100}{0 \cdot 80 \times 9,000} = 13 \cdot 3 \text{ lbs. of dry peat substance, or}$$

more than is contained in the peat.

The only possibility along this line is the employment of vacuum apparatus, but no such apparatus suitable for peat has yet been produced.

To eliminate the water content by mechanical pressure down to less than about 70%, has, with this apparatus, so far as tried, also been found practically impossible. The explanation is, according to Dr. A. Ekenberg, who has thoroughly studied this subject, the presence of a slimy hydrocellulose, which substance is produced from the cell residues of decaying plants by prolonged contact with water.

The principle of the Ekenberg wet carbonizing process is based on this point; the process aims at the destruction of this slimy substance and is obtained through the heating of the wet peat at a temperature of 155° to 180° Cent., with a corresponding pressure, in order to prevent the formation of steam and the loss of heat which such formation would involve. The process is further described in my report on Peat and Lignite in Europe, pages 160-170.

In cases where the peat contains a considerable amount of nitrogen, 1·5—3% (in dried sample), Dr. Ekenberg proposes to employ Mond gas-producers to furnish the gas required for the gas engines, as well as for the heating of the carbonizing ovens. In that case the nitrogen will be recovered as ammonium sulphate, the value of which will considerably decrease the cost of manufacture.

A plant with a capacity of 50 tons of briquettes per day of 24 hours will probably be in operation in Sweden during 1909.

The peat coking plant at Beuerberg, Germany, has, during the last year, been further improved by Mr. A. Ziegler. In order to produce a good coke, it is essential that the peat used should not contain more than 25% moisture. The air-dried peat is, however, very seldom uniform in moisture content, and the peat manufactured during the latter part of the summer very often contains more than 25% moisture. Mr. Ziegler overcomes this difficulty by employing the waste gases from the retorts for preliminary drying of the peat. The gases are mixed with air, so that a suitable temperature is obtained, and then passed through the peat in special drying chambers.

A new peat machine, with a large capacity, invented by Mr. A. Anrep, is at present being introduced in Europe, and the same inventor is also occupied with the construction of a mechanical excavator in combination with a peat plant of somewhat new design. The details of this plant are not as yet available. A new stump pulling apparatus has also been invented by Mr. Anrep.

For the manufacture of peat powder, in accordance with the process invented by Mr. H. Ekelund, of Sweden, an extensive plant was started last year and will probably be in full operation during the coming season.

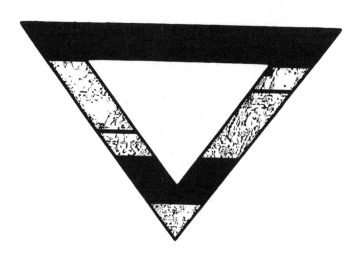

CPSIA information can be obtained
at www.ICGtesting.com
Printed in the USA
BVHW041431220219
540923BV00007B/345/P